Green Horizons

Navigating the Future of Our Planet with Sustainable Solutions

Dana Stokes

Table of Contents

Chapter 1

Introduction to Green Horizons

The Urgency of Environmental Action

The planet stands at a critical juncture, teetering on the edge of irreversible environmental damage. The urgency of environmental action has never been more pressing, as the consequences of inaction are becoming increasingly apparent. From the melting polar ice caps to the rampant deforestation of rainforests, the signs of ecological distress are all around us. The time to act is now, and the responsibility lies with each of us to make meaningful changes that can alter the course of our planet's future.

The evidence of environmental degradation is overwhelming. Climate change, driven by the relentless emission of greenhouse gases, is causing global temperatures to rise at an alarming rate. This warming effect is not just a distant threat; it is manifesting in the form of extreme weather events, such as hurricanes, droughts, and wildfires, which are becoming more frequent and severe. These events not only wreak havoc on natural ecosystems but also have devastating impacts on human communities, leading to loss of life, displacement, and economic hardship.

Biodiversity loss is another critical issue that underscores the urgency of environmental action. Species are disappearing at an unprecedented rate, with estimates suggesting that we are losing dozens of species every day. This loss of biodiversity is not just a tragedy for the natural world; it also poses a significant threat to human well-being. Ecosystems provide essential services, such as pollination, water purification, and carbon sequestration, which are vital for our survival. The decline in biodiversity undermines these services, putting our food security, health, and livelihoods at risk.

The degradation of natural resources further highlights the need for immediate action. Forests, which act as the lungs of our planet, are being cleared at an alarming rate to make way for agriculture and urban development. This deforestation not only contributes to climate change by releasing stored carbon dioxide but also destroys habitats and disrupts the delicate balance of ecosystems. Similarly, our oceans are under threat from overfishing, pollution, and acidification, which are depleting fish stocks and damaging marine habitats.

Water scarcity is another pressing issue that demands urgent attention. As populations grow and demand for water increases, many regions are facing severe shortages. Climate change exacerbates this problem by altering precipitation patterns and reducing the availability of freshwater resources. The consequences of water scarcity are far-reaching, affecting agriculture, industry, and human health. Without immediate action, billions of people could face water shortages in the coming decades, leading to conflicts and humanitarian crises.

The urgency of environmental action is further underscored by the interconnectedness of these issues. Climate change, biodiversity loss, resource depletion, and pollution are not isolated problems; they are deeply intertwined, each exacerbating the others. Addressing one issue without considering the others is insufficient. A holistic approach is needed, one that recognizes the complex interplay between human activities and the natural world.

The role of individuals, communities, and governments in driving environmental action cannot be overstated. Each of us has a part to play in reducing our environmental footprint and advocating for sustainable practices. Simple actions, such as reducing energy consumption, minimizing waste, and supporting sustainable products, can collectively make a significant impact. Communities can come together to implement local initiatives, such as tree planting, community gardens, and recycling programs, which not only benefit the environment but also foster a sense of shared responsibility and connection.

Governments have a crucial role in creating policies and regulations that promote environmental sustainability. This includes setting ambitious targets for reducing greenhouse gas emissions, protecting natural habitats, and investing in renewable energy sources. International cooperation is also essential, as environmental issues transcend national borders. Global agreements, such as the Paris Agreement, provide a framework for countries to work together towards common goals, but these commitments must be backed by concrete actions and accountability.

The private sector also has a significant role to play in driving environmental action. Businesses can adopt sustainable practices, such as reducing waste, improving energy efficiency, and sourcing materials responsibly. By prioritizing sustainability, companies can not only reduce their environmental impact but also gain a competitive advantage and build trust with consumers. Innovation and technology offer promising solutions to environmental challenges, from renewable energy technologies to sustainable agriculture practices. By investing in research and development, businesses can contribute to a more sustainable future.

Education and awareness are key components of driving environmental action. By understanding the urgency of the issues we face and the impact of our actions, individuals and communities can make informed decisions that support sustainability. Environmental education should be integrated into school curricula, empowering the next generation with the knowledge and skills needed to address these challenges. Public awareness campaigns can also play a vital role in changing attitudes and behaviors, encouraging people to adopt more sustainable lifestyles.

Understanding Sustainability

Sustainability is a concept that has gained significant traction in recent years, yet its true essence often remains elusive. At its core, sustainability is about meeting the needs of the present without compromising the ability of future generations to meet their own needs. This principle, rooted in the idea of balance, extends beyond environmental concerns to encompass economic and social dimensions as well. Understanding sustainability requires a holistic approach that considers the interconnectedness of these three pillars:

environmental integrity, economic viability, and social equity.

Environmental integrity is perhaps the most visible aspect of sustainability. It involves the responsible management of natural resources to ensure their availability for future generations. This means protecting ecosystems, conserving biodiversity, and reducing pollution. The health of our planet's ecosystems is crucial, as they provide essential services such as clean air and water, fertile soil, and climate regulation. By maintaining environmental integrity, we safeguard these services and ensure the resilience of natural systems in the face of challenges like climate change.

Economic viability is another critical component of sustainability. A sustainable economy is one that supports long-term prosperity without depleting natural resources or causing environmental harm. This involves rethinking traditional economic models that prioritize short-term gains over long-term stability. Sustainable economic practices include investing in renewable energy, promoting resource efficiency, and supporting local economies. By aligning economic activities with environmental goals, we can create a system that benefits both people and the planet.

Social equity is the third pillar of sustainability, emphasizing the importance of fairness and inclusivity. A sustainable society is one where all individuals have access to basic needs, such as food, water, shelter, and education, and where opportunities for growth and development are available to everyone. Social equity also involves addressing systemic inequalities and ensuring that marginalized communities are not disproportionately affected by environmental degradation or economic exploitation. By fostering social equity, we create a more just and resilient society.

Understanding sustainability also involves recognizing the interconnectedness of these three pillars. Environmental, economic, and social systems are deeply intertwined, and actions in one area can have ripple effects across the others. For example, investing in renewable energy not only reduces greenhouse gas emissions but also creates jobs and stimulates economic growth. Similarly, promoting social equity can lead to more sustainable resource management, as communities with access to education and opportunities are better equipped to make informed decisions about their environment.

The concept of sustainability is not static; it evolves as our understanding of the world and its challenges deepens. This dynamic nature requires adaptability and innovation. As new technologies and practices emerge, they offer opportunities to enhance sustainability efforts. However, it is essential to approach these innovations with a critical eye, ensuring that they align with the principles of sustainability and do not inadvertently cause harm.

One of the key challenges in understanding sustainability is the need for a long-term perspective. Many of the benefits of sustainable practices may not be immediately apparent, and the costs of inaction may only become evident over time. This requires a shift in mindset, from focusing on immediate gains to considering the broader implications of our actions. It also involves embracing uncertainty and being willing to experiment and learn from both successes and failures.

Education plays a vital role in fostering a deeper understanding of sustainability. By equipping individuals with the knowledge and skills needed to navigate complex environmental, economic, and social issues, we empower them to make informed decisions and take meaningful action. Education also helps to cultivate a sense of responsibility and stewardship, encouraging individuals to consider the impact of their choices on future generations.

Collaboration is another essential element of sustainability. No single entity can address the multifaceted challenges we face alone. Governments, businesses, communities, and individuals must work together to develop and implement sustainable solutions. This requires open communication, shared goals, and a willingness to compromise and adapt. By fostering collaboration, we can leverage diverse perspectives and expertise to create more effective and inclusive strategies.

Understanding sustainability also involves recognizing the importance of cultural and contextual factors. Different communities and regions may have unique challenges and opportunities, and sustainable solutions must be tailored to fit these specific contexts. This requires a deep understanding of local cultures, values, and practices, as well as a commitment to engaging with and listening to the voices of those most affected by sustainability issues.

The Role of Innovation in Environmental Solutions

Innovation has always been a driving force behind human progress, and its role in addressing environmental challenges is no exception. As the world grapples with issues such as climate change, resource depletion, and pollution, innovative solutions are essential to create a sustainable future. These solutions often emerge from the intersection of technology, science, and creativity, offering new ways to mitigate environmental impacts and enhance the resilience of natural systems.

One of the most promising areas of innovation in environmental solutions is renewable energy. The transition from fossil fuels to renewable sources such as solar, wind, and hydropower is crucial for reducing greenhouse gas emissions and combating climate change. Technological advancements have significantly improved the efficiency and affordability of renewable energy systems, making them more accessible to a broader range of users. For instance, solar panels have become more efficient at converting sunlight into electricity, while wind turbines have been designed to capture energy even at lower wind speeds. These innovations not only reduce our reliance on non-renewable resources but also create economic opportunities and jobs in the green energy sector.

Another area where innovation plays a critical role is in waste management and recycling. Traditional waste disposal methods, such as landfilling and incineration, pose significant environmental risks, including soil and water contamination and air pollution. Innovative approaches to waste management focus on reducing waste generation, increasing recycling rates, and recovering valuable materials. For example, advanced sorting technologies use artificial intelligence and robotics to separate recyclable materials more efficiently, while chemical recycling processes break down plastics into their original components for reuse. These innovations help to minimize the environmental footprint of waste and promote a circular economy, where resources are continuously reused and repurposed.

Water scarcity is another pressing environmental issue that demands innovative solutions. As populations grow and climate change alters precipitation patterns, access to clean and reliable water sources becomes increasingly challenging. Innovations in water management include technologies for desalination, water purification, and efficient irrigation. Desalination, which involves removing salt from seawater to produce fresh water, has become more energy-efficient and cost-effective, thanks to advances in membrane technology and energy recovery systems. Similarly, water purification technologies, such as advanced filtration and ultraviolet disinfection, ensure safe drinking water for communities worldwide. In agriculture, precision irrigation systems use sensors and data analytics to optimize water use, reducing waste and improving crop yields.

Biodiversity loss is another critical environmental challenge that requires innovative approaches. Habitat destruction, pollution, and climate change

threaten countless species and ecosystems, with far-
reaching consequences for human well-being.
Conservation efforts increasingly rely on innovative
tools and strategies to protect and restore
biodiversity. For example, satellite imagery and
remote sensing technologies enable scientists to
monitor ecosystems and track changes in real-time,
providing valuable data for conservation planning.
Additionally, genetic technologies, such as DNA
barcoding and gene editing, offer new ways to study
and preserve endangered species. These innovations
enhance our ability to understand and protect the
natural world, ensuring the survival of diverse
species and ecosystems.

Urbanization presents both challenges and
opportunities for environmental innovation. As
cities grow, they face increased pressure on
resources, infrastructure, and the environment.
Innovative urban planning and design can help
create sustainable cities that minimize environmental
impacts while enhancing quality of life. Green
building technologies, such as energy-efficient
materials and smart building systems, reduce energy
consumption and emissions in urban areas. Urban
green spaces, such as parks and green roofs, provide
essential ecosystem services, including air
purification, temperature regulation, and habitat for
wildlife. Additionally, sustainable transportation
solutions, such as electric vehicles and public transit

systems, reduce traffic congestion and pollution, contributing to cleaner and more livable cities.

The role of innovation in environmental solutions extends beyond technology to include social and behavioral change. Encouraging sustainable practices and lifestyles requires innovative approaches to education, communication, and policy. For example, gamification and digital platforms can engage individuals and communities in environmental initiatives, fostering a sense of responsibility and empowerment. Policies that incentivize sustainable behaviors, such as carbon pricing and subsidies for renewable energy, create an enabling environment for innovation and change. By integrating social and technological innovations, we can create a more sustainable and equitable future.

Collaboration is a key factor in driving innovation for environmental solutions. Partnerships between governments, businesses, academia, and civil society can accelerate the development and implementation of innovative solutions. Collaborative research and development initiatives, such as public-private partnerships and international collaborations, pool resources and expertise to tackle complex environmental challenges. By fostering a culture of collaboration and knowledge-sharing, we can harness the collective power of diverse stakeholders to create impactful and scalable solutions.

Innovation in environmental solutions is not without its challenges. The rapid pace of technological change can outstrip regulatory frameworks, leading to unintended consequences and ethical dilemmas. Ensuring that innovations align with sustainability goals requires careful consideration of their social, economic, and environmental impacts. Additionally, access to innovative solutions can be uneven, with marginalized communities often facing barriers to adoption. Addressing these challenges requires a commitment to inclusive and equitable innovation, ensuring that the benefits of environmental solutions are shared by all.

Global Environmental Challenges

The planet faces a multitude of environmental challenges that transcend borders and demand a collective response. These global issues, ranging from climate change to biodiversity loss, threaten the delicate balance of ecosystems and the well-being of all living beings. Understanding the complexity and interconnectedness of these challenges is crucial for developing effective strategies to address them.

Climate change stands as one of the most pressing global environmental challenges. Driven by the accumulation of greenhouse gases in the atmosphere, primarily from burning fossil fuels, climate change manifests in rising global temperatures, melting ice caps, and increasingly frequent and severe weather events. The impacts are far-reaching, affecting agriculture, water resources, and human health. Coastal communities face the threat of rising sea levels, while extreme weather events, such as hurricanes and droughts, disrupt lives and economies. Addressing climate change requires a multifaceted approach, including reducing emissions, transitioning to renewable energy, and enhancing resilience through adaptation measures.

Deforestation is another significant challenge with global implications. Forests play a vital role in regulating the Earth's climate by absorbing carbon dioxide and providing habitat for countless species. However, large-scale deforestation, driven by agriculture, logging, and urban expansion, contributes to carbon emissions and biodiversity loss. The Amazon rainforest, often referred to as the "lungs of the planet," has experienced alarming rates of deforestation, threatening its ability to sequester carbon and support diverse ecosystems. Efforts to combat deforestation include promoting sustainable land use practices, enforcing conservation policies, and supporting reforestation initiatives.

Biodiversity loss is a critical issue that affects the stability and resilience of ecosystems worldwide. Human activities, such as habitat destruction, pollution, and overexploitation of resources, have led to a dramatic decline in species populations and diversity. The loss of biodiversity undermines ecosystem services, such as pollination, water purification, and disease regulation, which are essential for human survival. Conservation efforts focus on protecting endangered species, restoring habitats, and promoting sustainable resource management. International agreements, such as the Convention on Biological Diversity, aim to halt biodiversity loss and ensure the sustainable use of natural resources.

Water scarcity is an escalating global challenge, exacerbated by population growth, climate change, and unsustainable water management practices. Access to clean and reliable water is essential for human health, agriculture, and industry. However, many regions face water stress, with demand outstripping supply. The depletion of aquifers, pollution of water sources, and inefficient irrigation practices contribute to the crisis. Addressing water scarcity requires integrated water management strategies, including improving water use efficiency, investing in infrastructure, and protecting freshwater ecosystems.

Pollution, in its various forms, poses a significant threat to the environment and human health. Air pollution, resulting from industrial emissions, vehicle exhaust, and burning of fossil fuels, contributes to respiratory diseases and climate change. Water pollution, caused by agricultural runoff, industrial discharges, and plastic waste, contaminates drinking water sources and harms aquatic life. Soil pollution, from pesticides and heavy metals, degrades land quality and affects food security. Tackling pollution involves implementing stricter regulations, promoting cleaner technologies, and raising awareness about sustainable practices.

The depletion of natural resources is another global challenge that demands urgent attention. The overconsumption of resources, such as minerals, fossil fuels, and freshwater, strains the planet's capacity to sustain human activities. Unsustainable resource extraction leads to environmental degradation, habitat destruction, and social conflicts. Transitioning to a circular economy, where resources are reused and recycled, can help reduce the pressure on natural resources and promote sustainable development.

Ocean health is a critical component of global environmental challenges. Oceans regulate the Earth's climate, provide food and livelihoods for billions of people, and support diverse marine ecosystems. However, human activities, such as overfishing, pollution, and climate change, threaten ocean health. Overfishing depletes fish stocks and disrupts marine food webs, while plastic pollution harms marine life and ecosystems. Ocean acidification, driven by increased carbon dioxide levels, affects coral reefs and shellfish populations. Protecting ocean health requires international cooperation, sustainable fisheries management, and efforts to reduce pollution and carbon emissions.

Urbanization presents both challenges and opportunities for addressing global environmental issues. As more people move to cities, the demand for resources, infrastructure, and services increases, leading to environmental pressures. Urban areas contribute significantly to greenhouse gas emissions, waste generation, and resource consumption. However, cities also offer opportunities for innovation and sustainable development. Implementing green infrastructure, promoting public transportation, and enhancing energy efficiency can help mitigate the environmental impacts of urbanization.

The intersection of environmental challenges with social and economic issues highlights the need for integrated solutions. Environmental degradation often exacerbates social inequalities, with marginalized communities disproportionately affected by pollution, resource scarcity, and climate impacts. Addressing these challenges requires a holistic approach that considers social, economic, and environmental dimensions. Policies that promote social equity, economic resilience, and environmental sustainability can help create a more just and sustainable world.

International cooperation is essential for tackling global environmental challenges. Many of these issues transcend national borders and require coordinated efforts to address them effectively. International agreements, such as the Paris Agreement on climate change and the Sustainable Development Goals, provide frameworks for collective action. Collaborative research, knowledge-sharing, and capacity-building initiatives can enhance global efforts to address environmental challenges.

The role of individuals and communities in addressing global environmental challenges cannot be underestimated. Grassroots movements, community-led initiatives, and individual actions contribute to environmental sustainability and resilience. By adopting sustainable practices, advocating for policy change, and raising awareness, individuals and communities can drive positive change and contribute to global efforts.

The Vision for a Sustainable Future

Imagining a sustainable future requires a shift in perspective, where the needs of the present are met without compromising the ability of future generations to meet their own. This vision encompasses a harmonious balance between economic growth, social equity, and environmental protection. It calls for innovative solutions, collaborative efforts, and a commitment to long-term thinking.

Central to this vision is the transformation of our energy systems. The reliance on fossil fuels has driven economic development for centuries, but it has also led to environmental degradation and climate change. Transitioning to renewable energy sources, such as solar, wind, and hydropower, is essential for reducing greenhouse gas emissions and mitigating climate impacts. This shift not only addresses environmental concerns but also presents economic opportunities. The renewable energy sector is a growing industry, creating jobs and stimulating technological advancements. Investing in clean energy infrastructure and research can accelerate this transition and pave the way for a sustainable energy future.

Sustainable agriculture is another cornerstone of this vision. The current agricultural practices, characterized by intensive resource use and chemical inputs, have led to soil degradation, water scarcity, and biodiversity loss. Embracing sustainable farming methods, such as agroecology, permaculture, and regenerative agriculture, can enhance soil health, conserve water, and promote biodiversity. These practices prioritize ecological balance and resilience, ensuring food security for future generations. Supporting local and organic food systems, reducing food waste, and promoting plant-based diets are additional strategies that contribute to sustainable agriculture.

Urban areas play a crucial role in shaping a sustainable future. As hubs of innovation and economic activity, cities have the potential to lead the way in sustainable development. Urban planning that prioritizes green spaces, public transportation, and energy-efficient buildings can reduce environmental impacts and enhance quality of life. Smart city technologies, such as energy-efficient lighting, waste management systems, and data-driven infrastructure, can optimize resource use and improve urban sustainability. Engaging communities in the planning process and fostering a sense of ownership and responsibility are key to creating resilient and sustainable cities.

The circular economy is a transformative concept that reimagines the traditional linear model of production and consumption. In a circular economy, resources are kept in use for as long as possible, waste is minimized, and products are designed for reuse, repair, and recycling. This approach reduces the pressure on natural resources, decreases waste, and fosters innovation. Businesses can adopt circular practices by designing products with longevity in mind, implementing take-back schemes, and using recycled materials. Consumers can contribute by choosing sustainable products, supporting companies with circular business models, and embracing a culture of sharing and repairing.

Education and awareness are fundamental to realizing a sustainable future. Empowering individuals with knowledge and skills enables them to make informed decisions and take meaningful action. Environmental education should be integrated into curricula at all levels, fostering a sense of stewardship and responsibility from an early age. Public awareness campaigns, community workshops, and online platforms can disseminate information and inspire collective action. By cultivating a culture of sustainability, individuals and communities can drive change and contribute to a more sustainable world.

Social equity is an integral component of a sustainable future. Environmental challenges often disproportionately affect marginalized communities, exacerbating existing inequalities. Ensuring that all individuals have access to clean air, water, and resources is essential for social justice and sustainability. Policies that promote inclusivity, equitable resource distribution, and community empowerment can address these disparities. Supporting grassroots movements, amplifying diverse voices, and fostering collaboration between stakeholders can create a more equitable and sustainable society.

Technological innovation is a powerful tool for advancing sustainability. From renewable energy technologies to sustainable transportation solutions, innovation can drive progress and address complex challenges. Research and development in fields such as biotechnology, materials science, and information technology can lead to breakthroughs that enhance sustainability. However, it is important to ensure that technological advancements are accessible and equitable, benefiting all segments of society. Collaborative efforts between governments, businesses, and academia can foster innovation and accelerate the transition to a sustainable future.

Policy and governance play a critical role in shaping a sustainable future. Effective policies can create an enabling environment for sustainable practices, incentivize innovation, and ensure accountability. Governments can implement regulations that promote renewable energy, sustainable agriculture, and resource efficiency. International cooperation is essential for addressing global challenges, such as climate change and biodiversity loss. Multilateral agreements, such as the Paris Agreement and the Sustainable Development Goals, provide frameworks for collective action and progress. Transparent and inclusive governance, with active participation from civil society, can ensure that policies are effective and equitable.

Chapter 2

The Science of Sustainability

Key Principles of Ecology

Ecology, the intricate web of interactions between organisms and their environment, is governed by a set of fundamental principles that shape the natural world. Understanding these principles is essential for grasping the complexity of ecosystems and the delicate balance that sustains life on Earth. These principles not only illuminate the interconnectedness of living organisms but also highlight the impact of human activities on ecological systems.

One of the foundational principles of ecology is the concept of interdependence. Every organism, from the smallest microbe to the largest mammal, relies on other organisms and the environment for survival. This interdependence is evident in the food chains and food webs that illustrate the flow of energy and nutrients through ecosystems. Producers, such as plants and algae, harness energy from the sun through photosynthesis, forming the base of the food chain. Herbivores consume these producers, while carnivores prey on herbivores, creating a complex network of interactions. Decomposers, such as fungi and bacteria, break down dead organic matter, recycling nutrients back into the ecosystem. This cyclical process ensures the continuity of life and the stability of ecosystems.

Another key principle is the concept of energy flow. Energy enters ecosystems primarily through sunlight, which is converted into chemical energy by producers. This energy is then transferred through the food chain, with each trophic level receiving only a fraction of the energy from the previous level. This loss of energy, primarily as heat, limits the number of trophic levels in an ecosystem and influences the structure and dynamics of communities. Understanding energy flow is crucial for managing ecosystems and ensuring their sustainability.

The principle of ecological niches highlights the role of species in ecosystems. Each species occupies a unique niche, defined by its habitat, diet, behavior, and interactions with other species. Niches reduce competition by allowing species to coexist and utilize different resources. When two species compete for the same niche, one may outcompete the other, leading to competitive exclusion. Alternatively, species may adapt and evolve to occupy different niches, a process known as niche differentiation. This principle underscores the importance of biodiversity and the role of species in maintaining ecosystem function.

Carrying capacity is a principle that describes the maximum population size an environment can sustain. It is determined by the availability of resources, such as food, water, and shelter, as well as environmental conditions. When a population exceeds its carrying capacity, resources become scarce, leading to competition, stress, and population decline. This principle is critical for understanding population dynamics and the impact of human activities on ecosystems. Overexploitation of resources, habitat destruction, and pollution can reduce carrying capacity, threatening the survival of species and the health of ecosystems.

The principle of succession describes the process of change in species composition and community structure over time. Succession occurs in response to disturbances, such as fires, storms, or human activities, that alter the environment. Primary succession begins in lifeless areas, such as bare rock or sand dunes, where pioneer species colonize and modify the environment, paving the way for other species. Secondary succession occurs in areas where a disturbance has removed existing vegetation but left the soil intact. Succession leads to the development of stable, mature communities, known as climax communities, that are adapted to the local environment. This principle highlights the resilience of ecosystems and their ability to recover from disturbances.

Biodiversity, the variety of life in all its forms, is a fundamental principle of ecology. It encompasses genetic diversity, species diversity, and ecosystem diversity, each contributing to the resilience and stability of ecosystems. High biodiversity enhances ecosystem function by providing a range of services, such as pollination, nutrient cycling, and climate regulation. It also increases the ability of ecosystems to withstand and recover from disturbances. Human activities, such as habitat destruction, pollution, and climate change, threaten biodiversity, with profound implications for ecosystem health and human well-being.

The principle of ecological interactions emphasizes the complex relationships between organisms and their environment. These interactions include competition, predation, mutualism, and parasitism, each influencing the distribution and abundance of species. Competition occurs when organisms vie for the same resources, while predation involves one organism consuming another. Mutualism is a cooperative interaction that benefits both species, such as pollinators and flowering plants. Parasitism involves one organism benefiting at the expense of another. These interactions shape community structure and drive evolutionary change, highlighting the dynamic nature of ecosystems. impact on ecosystems is a critical consideration in ecology. Human activities, such as deforestation, urbanization, and industrialization, have altered ecosystems and disrupted ecological processes. Pollution, climate change, and habitat fragmentation threaten biodiversity and ecosystem function. Understanding the principles of ecology is essential for mitigating these impacts and promoting sustainable practices. Conservation efforts, such as habitat restoration, protected areas, and sustainable resource management, aim to preserve biodiversity and maintain ecosystem health.

Climate Change and Its Impact

Climate change, a phenomenon characterized by long-term alterations in temperature, precipitation, and other atmospheric conditions, has emerged as one of the most pressing challenges of our time. Its impacts are far-reaching, affecting ecosystems, human societies, and the global economy. Understanding the causes and consequences of climate change is crucial for developing effective strategies to mitigate its effects and adapt to a changing world.

At the heart of climate change lies the increase in greenhouse gases (GHGs) in the atmosphere, primarily due to human activities. Carbon dioxide (CO2), methane (CH4), and nitrous oxide (N2O) are the main culprits, with CO2 being the most significant contributor. The burning of fossil fuels for energy, deforestation, and industrial processes release vast amounts of CO2, while agriculture and waste management contribute to methane and nitrous oxide emissions. These gases trap heat in the atmosphere, leading to a warming effect known as the greenhouse effect. This warming disrupts the delicate balance of Earth's climate system, resulting in a cascade of environmental changes.

One of the most visible impacts of climate change is the increase in global temperatures. The past century has witnessed a significant rise in average temperatures, with the last few decades being the warmest on record. This warming trend has profound implications for weather patterns, leading to more frequent and intense heatwaves, droughts, and storms. The increased frequency of extreme weather events poses significant risks to human health, agriculture, and infrastructure, challenging communities to adapt and build resilience.

The warming climate also affects the cryosphere, the frozen water part of the Earth system. Glaciers and ice sheets are melting at an alarming rate, contributing to rising sea levels. Coastal communities are particularly vulnerable, facing the threat of inundation and erosion. The loss of ice also has implications for global ocean circulation, which plays a crucial role in regulating climate. Changes in ocean currents can alter weather patterns and impact marine ecosystems, affecting biodiversity and fisheries.

Ecosystems around the world are experiencing shifts in response to climate change. Species are migrating to higher altitudes and latitudes in search of suitable habitats, leading to changes in community composition and interactions. Some species may face extinction if they cannot adapt or migrate quickly enough. These changes have cascading effects on ecosystem services, such as pollination, water purification, and carbon sequestration, which are vital for human well-being.

Climate change also exacerbates existing environmental issues, such as habitat loss, pollution, and overexploitation of resources. For instance, warmer temperatures can increase the prevalence of pests and diseases, affecting crops and livestock. This poses a threat to food security, particularly in regions already facing challenges related to poverty and resource scarcity. Water resources are also under pressure, with changing precipitation patterns leading to more frequent floods and droughts. These impacts highlight the interconnectedness of climate change with other environmental and social issues. societies are not only affected by the direct impacts of climate change but also by the measures taken to mitigate and adapt to it. Transitioning to a low-carbon economy requires significant changes in energy production, transportation, and land use. Renewable energy sources, such as solar and wind, offer promising alternatives to fossil fuels, but their implementation requires investment and infrastructure development. Energy efficiency and conservation measures can also play a crucial role in reducing emissions and minimizing the impact of climate change.

Adaptation strategies are essential for building resilience to the impacts of climate change. These strategies include improving infrastructure to withstand extreme weather events, developing early warning systems, and implementing sustainable land and water management practices. Community-based approaches that involve local stakeholders in decision-making processes can enhance the effectiveness of adaptation efforts and ensure that they are culturally and contextually appropriate.

International cooperation is vital for addressing the global challenge of climate change. The Paris Agreement, adopted in 2015, represents a significant step towards collective action, with countries committing to limit global warming to well below 2 degrees Celsius above pre-industrial levels. Achieving this goal requires ambitious efforts to reduce emissions, enhance carbon sinks, and support vulnerable communities in adapting to climate change. Financial and technological support from developed to developing countries is crucial for enabling these efforts and ensuring a just transition.

Public awareness and engagement are also critical components of climate action. Educating individuals and communities about the causes and impacts of climate change can empower them to make informed decisions and take action. Simple lifestyle changes, such as reducing energy consumption, minimizing waste, and supporting sustainable products, can collectively make a significant difference. Advocacy and participation in policy processes can also drive systemic change and hold governments and corporations accountable for their actions.

The challenge of climate change is immense, but it also presents opportunities for innovation and transformation. By embracing sustainable practices and technologies, societies can create a more resilient and equitable future. The transition to a low-carbon economy can stimulate economic growth, create jobs, and improve public health. Protecting and restoring natural ecosystems can enhance

biodiversity and provide valuable ecosystem services.
Ultimately, addressing climate change requires a
holistic approach that considers environmental,
social, and economic dimensions, recognizing that
the health of the planet and the well-being of its
inhabitants are inextricably linked.

Biodiversity and Ecosystem Services

Biodiversity, the variety of life on Earth,
encompasses the multitude of species, genetic
variations, and ecosystems that make up our planet.
It is the foundation upon which ecosystems function
and provide essential services that sustain human
life. These ecosystem services include provisioning,
regulating, supporting, and cultural benefits that are
vital for our survival and well-being. However,
biodiversity is under threat from human activities,
and its decline poses significant risks to the stability
and resilience of ecosystems.

The richness of biodiversity is evident in the myriad
of species that inhabit different ecosystems, from the
lush rainforests teeming with life to the arid deserts
where only the most resilient organisms thrive. Each
species plays a unique role in its ecosystem,
contributing to processes such as pollination,
nutrient cycling, and soil formation. The loss of even

a single species can disrupt these processes, leading to cascading effects that impact the entire ecosystem. For example, the decline of pollinators like bees and butterflies can affect crop yields and food security, highlighting the interconnectedness of biodiversity and human livelihoods.

Genetic diversity, a component of biodiversity, is crucial for the adaptability and resilience of species. It enables populations to withstand environmental changes, diseases, and other threats. In agriculture, genetic diversity in crops and livestock is essential for breeding programs that develop varieties resistant to pests, diseases, and climate change. The erosion of genetic diversity, often due to monoculture practices and habitat destruction, reduces the ability of species to adapt and survive, making them more vulnerable to extinction.

Ecosystems, the dynamic complexes of plant, animal, and microorganism communities interacting with their physical environment, provide a range of services that are indispensable to human societies. Provisioning services include the supply of food, fresh water, timber, and medicinal resources. Many of the medicines we rely on today are derived from plants and animals, underscoring the importance of conserving biodiversity for future discoveries. Regulating services, such as climate regulation, water purification, and disease control, help maintain environmental stability and protect human health. Wetlands, for instance, act as natural water filters, removing pollutants and improving water quality.

Supporting services are the underlying processes that enable ecosystems to function and provide other services. These include nutrient cycling, soil formation, and primary production. Healthy soils, rich in organic matter and microorganisms, support plant growth and agricultural productivity. The loss of supporting services can lead to soil degradation, reduced crop yields, and increased vulnerability to natural disasters. Cultural services, the non-material benefits people obtain from ecosystems, include recreation, spiritual enrichment, and aesthetic enjoyment. Natural landscapes and biodiversity inspire art, literature, and cultural traditions, contributing to our sense of identity and well-being.

The decline of biodiversity and the degradation of ecosystem services are driven by various factors, including habitat loss, pollution, overexploitation, invasive species, and climate change. Deforestation, urbanization, and agricultural expansion lead to habitat fragmentation and destruction, reducing the available space for species to thrive. Pollution from industrial activities, agriculture, and waste disposal contaminates air, water, and soil, harming wildlife and ecosystems. Overexploitation of resources, such as overfishing and illegal wildlife trade, depletes populations and disrupts ecological balance.

Invasive species, introduced intentionally or accidentally, can outcompete native species for resources, leading to declines in biodiversity. Climate change exacerbates these threats by altering temperature and precipitation patterns, affecting species' distribution and behavior. Coral reefs, for example, are highly sensitive to temperature changes, and rising sea temperatures have led to widespread coral bleaching and mortality. The loss of coral reefs impacts marine biodiversity and the livelihoods of communities that depend on them for food and tourism.

Conserving biodiversity and maintaining ecosystem services require concerted efforts at local, national, and global levels. Protected areas, such as national parks and wildlife reserves, play a crucial role in safeguarding habitats and species. Effective management of these areas involves monitoring biodiversity, controlling invasive species, and engaging local communities in conservation efforts. Restoration projects, such as reforestation and wetland rehabilitation, can help restore degraded ecosystems and enhance their capacity to provide services.

Sustainable land and resource management practices are essential for balancing human needs with environmental conservation. Agroecology, an approach that integrates ecological principles into agricultural practices, promotes biodiversity and ecosystem services while enhancing food security. It involves diversifying crops, reducing chemical inputs, and conserving natural habitats within agricultural landscapes. Community-based conservation initiatives empower local people to manage and benefit from natural resources sustainably, fostering a sense of stewardship and ownership.

International agreements and policies, such as the Convention on Biological Diversity and the United Nations Sustainable Development Goals, provide frameworks for global cooperation in biodiversity conservation. These agreements emphasize the need for equitable sharing of benefits arising from the use of genetic resources and the importance of integrating biodiversity considerations into development planning. Financial and technical support from developed to developing countries is crucial for implementing conservation strategies and addressing the root causes of biodiversity loss.

Public awareness and education are vital for fostering a culture of conservation and encouraging responsible behavior. By understanding the value of biodiversity and ecosystem services, individuals can make informed choices that contribute to conservation efforts. Simple actions, such as reducing waste, supporting sustainable products, and participating in citizen science projects, can collectively make a significant impact. Advocacy and engagement in policy processes can also drive systemic change and hold decision-makers accountable for their actions.

The Carbon Cycle and Its Importance

The carbon cycle is a fundamental component of Earth's system, intricately woven into the fabric of life and the environment. It is a complex network of processes that circulate carbon among the atmosphere, oceans, soil, and living organisms. Understanding the carbon cycle is crucial, as it plays a pivotal role in regulating Earth's climate, supporting life, and maintaining ecological balance. The cycle's significance extends beyond natural processes, influencing human activities and global policies aimed at mitigating climate change.

Carbon, a versatile element, is the building block of life. It forms the backbone of organic molecules, including carbohydrates, proteins, and nucleic acids, which are essential for the structure and function of living organisms. The carbon cycle begins with the fixation of atmospheric carbon dioxide (CO_2)

through photosynthesis, a process carried out by plants, algae, and certain bacteria. These organisms convert CO2 into organic matter, using sunlight as an energy source. This transformation not only provides the foundation for food chains but also acts as a natural mechanism for removing CO2 from the atmosphere.

Once carbon is incorporated into organic matter, it enters the food web. Herbivores consume plants, assimilating carbon into their bodies, while carnivores obtain carbon by eating herbivores. Decomposers, such as fungi and bacteria, break down dead organic matter, releasing carbon back into the atmosphere as CO2 through respiration. This continuous exchange of carbon between living organisms and the environment is a vital component of the carbon cycle, ensuring the flow of energy and nutrients within ecosystems.

The oceans, covering over 70% of Earth's surface, are a major reservoir of carbon. They absorb CO2 from the atmosphere, a process facilitated by the solubility of CO2 in water. Once dissolved, CO2 can be used by marine organisms, such as phytoplankton, for photosynthesis. The carbon is then transferred through the marine food web, eventually reaching the deep ocean as organic matter sinks. This process, known as the biological pump, sequesters carbon in the ocean depths for centuries, playing a critical role in regulating atmospheric CO2 levels.

Soil is another significant carbon reservoir, storing more carbon than the atmosphere and living organisms combined. Organic matter, derived from plant and animal residues, accumulates in the soil, where it is decomposed by microorganisms. The rate of decomposition is influenced by factors such as temperature, moisture, and soil composition. In some cases, carbon can be stored in the soil for long periods, contributing to soil fertility and structure. However, land-use changes, such as deforestation and agriculture, can disrupt this balance, releasing stored carbon into the atmosphere and exacerbating climate change. activities have profoundly impacted the carbon cycle, primarily through the burning of fossil fuels and land-use changes. The combustion of coal, oil, and natural gas releases vast amounts of CO2, a greenhouse gas that traps heat in the atmosphere and contributes to global warming. Deforestation, driven by agriculture and urban expansion, reduces the capacity of forests to absorb CO2, further increasing atmospheric carbon levels. These activities have altered the natural carbon cycle, leading to a rise in global temperatures and associated climate impacts.

The importance of the carbon cycle extends to its role in climate regulation. Greenhouse gases, including CO2, methane, and nitrous oxide, are integral to maintaining Earth's energy balance. They allow sunlight to enter the atmosphere while trapping some of the outgoing heat, creating a natural greenhouse effect that keeps the planet warm enough to support life. However, the excessive accumulation of greenhouse gases, driven by human activities, intensifies this effect, leading to climate change. Understanding the carbon cycle is essential for developing strategies to mitigate these impacts and stabilize the climate.

Efforts to address climate change often focus on reducing carbon emissions and enhancing carbon sinks. Renewable energy sources, such as solar, wind, and hydropower, offer alternatives to fossil fuels, reducing CO2 emissions. Energy efficiency measures, carbon capture and storage technologies, and reforestation projects also contribute to emission reductions. Protecting and restoring natural ecosystems, such as forests, wetlands, and grasslands, enhances their capacity to absorb and store carbon, providing a natural solution to climate mitigation.

International agreements, such as the Paris Agreement, aim to limit global warming by reducing greenhouse gas emissions and promoting sustainable development. These agreements emphasize the importance of the carbon cycle in achieving climate goals and encourage countries to implement policies that support carbon management. Collaboration among governments, businesses, and communities is crucial for achieving these objectives and ensuring a sustainable future.

Public awareness and education play a vital role in promoting understanding of the carbon cycle and its importance. By recognizing the interconnectedness of human activities and natural processes, individuals can make informed choices that contribute to carbon management. Simple actions, such as reducing energy consumption, supporting sustainable products, and participating in conservation efforts, can collectively make a significant impact. Advocacy and engagement in policy processes can also drive systemic change and hold decision-makers accountable for their actions.

Renewable Resources vs Nonrenewable Resources

The distinction between renewable and nonrenewable resources is a cornerstone of understanding how we interact with the environment and manage the Earth's natural wealth. These resources form the backbone of our economies, influence our lifestyles, and shape our future. Grasping the differences between them is essential for making informed decisions about resource management, sustainability, and environmental impact.

Renewable resources are those that can be replenished naturally over short periods. They include solar energy, wind, water (hydropower), biomass, and geothermal energy. These resources are often seen as sustainable alternatives to traditional energy sources because they are abundant and have a lower environmental impact. Solar energy, for instance, harnesses the power of the sun, which is inexhaustible on a human timescale. Photovoltaic cells and solar panels convert sunlight into electricity, providing a clean and renewable energy source. Wind energy, captured through turbines, is another example of a renewable resource that offers a sustainable solution to energy needs.

Water, in the form of hydropower, is a renewable resource that has been used for centuries. By harnessing the kinetic energy of flowing water, we can generate electricity without emitting greenhouse gases. Biomass, derived from organic materials such as plants and animal waste, can be converted into biofuels or used directly for heating and electricity generation. Geothermal energy taps into the Earth's internal heat, providing a consistent and reliable energy source. These renewable resources offer the potential to reduce our reliance on fossil fuels and mitigate the impacts of climate change.

Nonrenewable resources, on the other hand, are finite and cannot be replenished within a human lifespan. They include fossil fuels such as coal, oil, and natural gas, as well as minerals and metals like gold, silver, and iron. The extraction and consumption of these resources have significant environmental and economic implications. Fossil fuels, for example, are the primary source of energy for many countries, but their combustion releases carbon dioxide and other pollutants, contributing to air pollution and climate change. The extraction process itself can lead to habitat destruction, water contamination, and other environmental issues.

The finite nature of nonrenewable resources poses a challenge for sustainable development. As these resources become scarcer, their extraction becomes more difficult and costly, leading to economic and geopolitical tensions. The reliance on nonrenewable resources also raises concerns about energy security, as countries compete for access to dwindling supplies. This has prompted a global shift towards renewable energy sources, which offer a more sustainable and secure energy future.

The transition from nonrenewable to renewable resources is not without challenges. Renewable energy technologies often require significant upfront investment and infrastructure development. The intermittent nature of some renewable resources, such as solar and wind, necessitates the development of energy storage solutions and grid management systems to ensure a stable energy supply. Additionally, the production and disposal of renewable energy technologies can have environmental impacts, such as the mining of rare earth metals for solar panels and wind turbines.

Despite these challenges, the benefits of renewable resources are compelling. They offer a pathway to reducing greenhouse gas emissions, decreasing air pollution, and minimizing the environmental footprint of energy production. The shift towards renewable resources also presents economic opportunities, as the renewable energy sector is a growing industry that creates jobs and stimulates innovation. Governments, businesses, and communities are increasingly recognizing the importance of investing in renewable energy and implementing policies that support the transition.

Public awareness and education are crucial in promoting the adoption of renewable resources. By understanding the environmental and economic benefits of renewable energy, individuals can make informed choices about their energy consumption and support policies that encourage sustainable practices. Simple actions, such as reducing energy use, investing in energy-efficient appliances, and supporting renewable energy initiatives, can collectively make a significant impact.

The role of technology and innovation in advancing renewable resources cannot be overstated. Advances in solar panel efficiency, wind turbine design, and energy storage solutions are making renewable energy more accessible and cost-effective. Research and development in areas such as bioenergy, hydrogen fuel cells, and smart grid technology are paving the way for a more sustainable energy future. Collaboration between governments, research institutions, and the private sector is essential for driving innovation and overcoming the challenges associated with renewable energy.

The transition to renewable resources is a global endeavor that requires cooperation and commitment from all sectors of society. International agreements, such as the Paris Agreement, emphasize the importance of reducing greenhouse gas emissions and promoting sustainable development. These agreements encourage countries to implement policies that support renewable energy and resource management, fostering a collaborative approach to addressing environmental challenges.

The distinction between renewable and nonrenewable resources highlights the need for a balanced approach to resource management. While nonrenewable resources have historically driven economic growth and development, their environmental and economic limitations necessitate a shift towards more sustainable alternatives. Renewable resources offer a viable solution, providing a pathway to a cleaner, more secure, and sustainable energy future.

Chapter 3

Innovative Technologies for a Greener Planet

Breakthroughs in Renewable Energy

The landscape of renewable energy has been transformed by a series of groundbreaking innovations that have redefined how we harness, store, and utilize natural resources. These breakthroughs are not only reshaping the energy sector but also paving the way for a more sustainable and resilient future. As the world grapples with the pressing challenges of climate change and resource depletion, the importance of these advancements cannot be overstated.

One of the most significant breakthroughs in renewable energy is the dramatic improvement in solar panel efficiency. Traditional photovoltaic cells have long been the cornerstone of solar energy, but recent developments have pushed their efficiency to new heights. Researchers have introduced perovskite solar cells, which offer a promising alternative to conventional silicon-based cells. These new materials are not only cheaper to produce but also boast higher efficiency rates, making solar energy more accessible and affordable. The potential to integrate perovskite cells with existing technologies could revolutionize the solar industry, enabling widespread adoption and reducing reliance on fossil fuels.

Wind energy has also seen remarkable advancements, particularly in the design and efficiency of wind turbines. Modern turbines are now capable of capturing wind energy at lower speeds, expanding the geographical areas where wind power can be effectively harnessed. Innovations such as floating wind farms have opened up new possibilities for offshore wind energy, allowing turbines to be placed in deeper waters where wind speeds are higher and more consistent. These developments have significantly increased the potential for wind energy to contribute to the global energy mix, providing a reliable and sustainable power source.

Energy storage is another critical area where breakthroughs have had a profound impact. The intermittent nature of renewable energy sources like solar and wind necessitates efficient storage solutions to ensure a stable energy supply. Advances in battery technology, particularly in lithium-ion and solid-state batteries, have greatly enhanced energy storage capabilities. These batteries offer higher energy densities, longer lifespans, and improved safety features, making them ideal for both grid-scale storage and electric vehicles. The development of flow batteries, which use liquid electrolytes to store energy, presents another promising avenue for large-scale energy storage, offering the potential for longer discharge times and greater scalability.

Hydrogen fuel cells represent a breakthrough in clean energy technology, offering a versatile and sustainable solution for various applications. By converting hydrogen gas into electricity, these fuel cells produce only water and heat as byproducts, making them an environmentally friendly alternative to traditional combustion engines. Recent advancements in hydrogen production, particularly through electrolysis powered by renewable energy, have made it more feasible to produce green hydrogen at scale. This development has significant implications for sectors such as transportation and industry, where hydrogen fuel cells can replace fossil fuels and reduce carbon emissions.

Geothermal energy, often overlooked in discussions of renewable energy, has also experienced notable advancements. Enhanced geothermal systems (EGS) have the potential to unlock vast amounts of geothermal energy by artificially creating reservoirs in hot, dry rock formations. This technology expands the geographical reach of geothermal energy, making it accessible in regions where traditional geothermal resources are not available. The ability to tap into the Earth's internal heat on a larger scale could provide a consistent and reliable energy source, complementing other renewable technologies.

The integration of smart grid technology represents a breakthrough in how renewable energy is managed and distributed. Smart grids use digital communication and automation to optimize the flow of electricity, allowing for more efficient use of renewable resources. By incorporating real-time data and advanced analytics, smart grids can balance supply and demand, reduce energy waste, and enhance grid reliability. This technology also enables greater integration of distributed energy resources, such as rooftop solar panels and small-scale wind turbines, empowering consumers to become active participants in the energy system.

Bioenergy, derived from organic materials, has seen significant advancements in recent years. Innovations in biofuel production, such as the development of advanced biofuels from non-food crops and agricultural waste, have improved the sustainability and efficiency of bioenergy. These biofuels offer a renewable alternative to fossil fuels, particularly in sectors like aviation and shipping, where electrification is challenging. The use of algae as a feedstock for biofuel production presents another promising avenue, as algae can be grown in a variety of environments and have a high oil content, making them an efficient source of bioenergy.

The role of policy and regulation in driving breakthroughs in renewable energy cannot be overlooked. Governments around the world are implementing policies that incentivize research and development, support the deployment of renewable technologies, and create favorable market conditions for clean energy. These policies, coupled with international agreements and commitments to reduce carbon emissions, are accelerating the transition to a renewable energy future.

Public awareness and engagement are also crucial in fostering innovation and adoption of renewable energy technologies. As individuals become more informed about the environmental and economic benefits of renewable energy, they are more likely to support policies and initiatives that promote sustainable practices. Community-based renewable energy projects, such as solar cooperatives and local wind farms, empower individuals to take an active role in the energy transition, fostering a sense of ownership and responsibility.

The breakthroughs in renewable energy are a testament to human ingenuity and the relentless pursuit of a sustainable future. These innovations are not only transforming the energy landscape but also offering hope for a world where clean, affordable, and reliable energy is accessible to all. As we continue to push the boundaries of what is possible, the potential for renewable energy to drive economic growth, enhance energy security, and mitigate the impacts of climate change becomes increasingly evident.

Advances in Waste Management

Waste management has undergone a remarkable transformation in recent years, driven by technological advancements and a growing awareness of environmental sustainability. As urban populations swell and consumption patterns evolve, the challenge of managing waste efficiently and responsibly has become more pressing than ever. The innovations in this field are not only addressing the immediate concerns of waste disposal but are also contributing to a broader vision of a circular economy, where waste is minimized, and resources are reused and recycled.

One of the most significant advances in waste management is the development of smart waste collection systems. These systems leverage sensors and data analytics to optimize waste collection routes, reducing fuel consumption and emissions. By monitoring the fill levels of waste bins in real-time, municipalities can deploy collection vehicles only when necessary, leading to more efficient operations and cost savings. This technology not only enhances the efficiency of waste collection but also reduces the environmental impact associated with traditional waste management practices.

Recycling technologies have also seen substantial improvements, particularly in the sorting and processing of materials. Automated sorting systems, equipped with advanced sensors and artificial intelligence, can now accurately identify and separate different types of recyclables, such as plastics, metals, and paper. This precision reduces contamination rates and increases the quality of recycled materials, making them more valuable and easier to reintroduce into the manufacturing process. The development of chemical recycling methods, which break down plastics into their basic chemical components, offers a promising solution for dealing with complex and mixed plastic waste that is difficult to recycle mechanically.

Composting has emerged as a vital component of waste management, particularly for organic waste. Innovations in composting technology have made it possible to process organic waste more efficiently and at a larger scale. Aerobic digesters, for example, accelerate the decomposition process by maintaining optimal conditions for microbial activity, resulting in high-quality compost that can be used to enrich soil and support sustainable agriculture. Community composting initiatives are also gaining traction, empowering individuals and local organizations to manage organic waste locally and reduce the burden on municipal waste systems.

The concept of waste-to-energy has gained momentum as a viable solution for reducing landfill waste and generating renewable energy. Modern waste-to-energy plants use advanced combustion and gasification technologies to convert non-recyclable waste into electricity and heat. These facilities are designed to minimize emissions and maximize energy recovery, providing a sustainable alternative to traditional waste disposal methods. The integration of carbon capture and storage technologies further enhances the environmental benefits of waste-to-energy, capturing carbon dioxide emissions and preventing them from entering the atmosphere.

E-waste, or electronic waste, presents a unique challenge due to the rapid pace of technological advancement and the short lifespan of electronic devices. Advances in e-waste recycling have focused on recovering valuable materials, such as precious metals and rare earth elements, from discarded electronics. Innovative processes, such as hydrometallurgical and biotechnological methods, offer efficient and environmentally friendly ways to extract these materials, reducing the need for virgin resource extraction and minimizing the environmental impact of e-waste.

The role of policy and regulation in driving advances in waste management cannot be understated. Governments worldwide are implementing policies that promote waste reduction, recycling, and sustainable waste management practices. Extended producer responsibility (EPR) schemes, for example, hold manufacturers accountable for the end-of-life management of their products, incentivizing them to design products that are easier to recycle and have a lower environmental impact. These policies, coupled with public awareness campaigns, are fostering a culture of sustainability and encouraging individuals and businesses to adopt more responsible waste management practices.

Public engagement and education are crucial components of effective waste management. As individuals become more informed about the environmental and economic benefits of sustainable waste practices, they are more likely to participate in recycling programs and support initiatives that promote waste reduction. Community-based programs, such as zero-waste events and educational workshops, empower individuals to take an active role in waste management, fostering a sense of responsibility and stewardship.

The shift towards a circular economy represents a fundamental change in how we view waste. Instead of seeing waste as an end product, the circular economy envisions waste as a resource that can be reused, recycled, or repurposed. This approach not only reduces the environmental impact of waste but also creates economic opportunities by generating value from materials that would otherwise be discarded. Advances in waste management are critical to realizing this vision, providing the tools and technologies needed to close the loop and create a more sustainable and resilient future.

Smart Agriculture and Food Security

The intersection of smart agriculture and food security represents a pivotal development in addressing the global challenges of feeding a growing population while ensuring sustainable agricultural practices. As the world grapples with climate change, resource scarcity, and population growth, the need for innovative solutions in agriculture has never been more urgent. Smart agriculture, characterized by the integration of advanced technologies into farming practices, offers a promising pathway to enhance food security and promote environmental sustainability.

At the heart of smart agriculture lies precision farming, a method that utilizes data-driven insights to optimize crop production. By employing sensors, GPS technology, and data analytics, farmers can monitor soil conditions, weather patterns, and crop health in real-time. This information allows for precise application of water, fertilizers, and pesticides, reducing waste and minimizing environmental impact. Precision farming not only increases crop yields but also conserves resources, making it a cornerstone of sustainable agriculture.

The role of the Internet of Things (IoT) in smart agriculture cannot be overstated. IoT devices, such as soil moisture sensors and weather stations, provide farmers with continuous data streams that inform decision-making processes. These devices enable farmers to implement site-specific management practices, tailoring interventions to the unique needs of each field. The result is a more efficient use of inputs and a reduction in the environmental footprint of agriculture. Moreover, IoT technology facilitates remote monitoring and management, allowing farmers to oversee their operations from anywhere in the world.

Drones have emerged as a valuable tool in the smart agriculture arsenal, offering a bird's-eye view of fields and providing critical insights into crop health and growth patterns. Equipped with high-resolution cameras and multispectral sensors, drones can capture detailed images that reveal variations in plant health, soil conditions, and pest infestations. This aerial perspective enables farmers to identify issues early and take targeted action, preventing crop losses and improving overall productivity. The use of drones in agriculture exemplifies the shift towards more proactive and informed farming practices.

The integration of blockchain technology into agriculture is transforming supply chain management and enhancing food security. Blockchain provides a transparent and immutable record of transactions, ensuring traceability and accountability throughout the supply chain. This technology enables consumers to verify the origin and quality of their food, fostering trust and confidence in the food system. For farmers, blockchain offers a platform to showcase sustainable practices and connect directly with consumers, opening new markets and opportunities for value-added products.

Climate-smart agriculture is an essential component of the broader smart agriculture framework, addressing the dual challenges of climate change and food security. This approach emphasizes the adoption of practices that increase resilience to climate variability while reducing greenhouse gas emissions. Techniques such as agroforestry, conservation tillage, and integrated pest management contribute to climate-smart agriculture by enhancing soil health, conserving biodiversity, and reducing reliance on chemical inputs. By aligning agricultural practices with environmental goals, climate-smart agriculture supports long-term food security and ecosystem health.

The potential of smart agriculture to improve food security extends beyond crop production to include livestock management. Precision livestock farming employs technologies such as wearable sensors and automated feeding systems to monitor animal health and optimize production. These innovations enable farmers to detect health issues early, improve animal welfare, and increase efficiency in feed and water use. By enhancing the productivity and sustainability of livestock systems, smart agriculture contributes to a more secure and resilient food supply.

Education and training are critical to the successful implementation of smart agriculture. Farmers must be equipped with the knowledge and skills to leverage new technologies and adapt to changing conditions. Extension services, agricultural colleges, and industry partnerships play a vital role in providing training and support to farmers, ensuring they can fully realize the benefits of smart agriculture. By fostering a culture of innovation and continuous learning, the agricultural sector can remain agile and responsive to emerging challenges.

The social and economic implications of smart agriculture are profound, offering opportunities to improve livelihoods and reduce poverty in rural communities. By increasing productivity and efficiency, smart agriculture can enhance farm profitability and create new employment opportunities in technology and service sectors. Additionally, the focus on sustainable practices and resource conservation aligns with broader goals of social equity and environmental stewardship, contributing to a more just and sustainable food system.

Collaboration and partnerships are essential to advancing smart agriculture and achieving food security. Governments, research institutions, private companies, and non-governmental organizations must work together to develop and disseminate innovative solutions. Public-private partnerships can facilitate the transfer of technology and knowledge, ensuring that farmers have access to the tools and resources they need to succeed. By fostering collaboration across sectors and borders, the global community can address the complex challenges of food security and build a more resilient agricultural system.

Water Conservation Technologies

Water conservation technologies have become increasingly vital as the world faces mounting pressures from climate change, population growth, and dwindling freshwater resources. These technologies offer innovative solutions to manage water more efficiently, ensuring that this precious resource is available for future generations. By adopting water conservation technologies, individuals, communities, and industries can significantly reduce water waste and promote sustainable water use.

One of the most effective water conservation technologies is drip irrigation, a method that delivers water directly to the roots of plants through a network of tubes and emitters. This system minimizes evaporation and runoff, ensuring that water is used efficiently and effectively. Drip irrigation is particularly beneficial in arid regions where water scarcity is a pressing concern. By providing a steady supply of moisture to crops, this technology enhances agricultural productivity while conserving water resources.

Rainwater harvesting is another crucial technology that captures and stores rainwater for later use. This practice involves collecting rainwater from rooftops, surfaces, or other catchment areas and storing it in tanks or reservoirs. The harvested water can be used for various purposes, such as irrigation, flushing toilets, or even drinking, depending on the level of treatment it receives. Rainwater harvesting reduces reliance on municipal water supplies and helps mitigate the impact of droughts and water shortages.

Greywater recycling is an innovative approach to water conservation that involves reusing wastewater from sinks, showers, and washing machines for non-potable purposes. By treating and repurposing greywater, households and businesses can significantly reduce their freshwater consumption. This technology not only conserves water but also reduces the burden on wastewater treatment facilities, contributing to more sustainable water management practices.

Smart water meters are transforming the way we monitor and manage water usage. These devices provide real-time data on water consumption, allowing users to identify patterns, detect leaks, and make informed decisions about water use. By offering insights into water usage, smart meters empower individuals and organizations to adopt more efficient practices and reduce waste. The data collected by smart meters can also inform policy decisions and support efforts to promote water conservation at a broader scale.

Low-flow fixtures, such as faucets, showerheads, and toilets, are simple yet effective technologies that reduce water consumption without sacrificing performance. These fixtures are designed to use less water by incorporating aerators, pressure regulators, or dual-flush mechanisms. By installing low-flow fixtures, households and businesses can achieve significant water savings, contributing to more sustainable water use.

Desalination technology offers a solution to water scarcity by converting seawater into freshwater. This process involves removing salt and other impurities from seawater through methods such as reverse osmosis or distillation. While desalination can provide a reliable source of freshwater, it is energy-intensive and can have environmental impacts. Advances in desalination technology are focused on improving efficiency and reducing costs, making it a more viable option for addressing water shortages in coastal regions.

Constructed wetlands are engineered systems that mimic natural wetlands to treat wastewater and stormwater. These systems use vegetation, soil, and microorganisms to remove pollutants and improve water quality. Constructed wetlands offer a sustainable and cost-effective solution for wastewater treatment, while also providing habitat for wildlife and enhancing biodiversity. By integrating constructed wetlands into urban and rural landscapes, communities can improve water management and promote ecological health.

Water-efficient landscaping, also known as xeriscaping, is a design approach that emphasizes the use of drought-tolerant plants and efficient irrigation practices. By selecting native or adapted plants that require minimal water, homeowners and landscapers can create beautiful and sustainable outdoor spaces. Water-efficient landscaping reduces the need for irrigation, conserves water, and supports local ecosystems, making it an essential component of water conservation efforts.

Leak detection and repair technologies play a critical role in preventing water loss and ensuring the integrity of water distribution systems. Advanced sensors and monitoring systems can detect leaks in pipelines and infrastructure, allowing for timely repairs and reducing water waste. By addressing leaks promptly, utilities and municipalities can conserve water, reduce costs, and improve the reliability of water supply systems.

Public awareness and education are fundamental to the success of water conservation technologies. By raising awareness about the importance of water conservation and the available technologies, individuals and communities can be motivated to adopt more sustainable practices. Educational programs, workshops, and outreach initiatives can provide valuable information and resources, empowering people to make informed decisions about water use.

The implementation of water conservation technologies requires collaboration and commitment from various stakeholders, including governments, industries, and communities. Policymakers play a crucial role in creating an enabling environment for the adoption of these technologies through incentives, regulations, and support for research and development. Industries can contribute by investing in water-efficient technologies and practices, while communities can advocate for sustainable water management and participate in conservation initiatives.

The Role of Artificial Intelligence in Sustainability

Artificial intelligence (AI) has emerged as a transformative force in the quest for sustainability, offering innovative solutions to some of the most pressing environmental challenges of our time. By harnessing the power of AI, we can optimize resource use, reduce waste, and promote more sustainable practices across various sectors. The integration of AI into sustainability efforts is not just a technological advancement; it represents a paradigm shift in how we approach environmental stewardship.

One of the most significant contributions of AI to sustainability is its ability to analyze vast amounts of data quickly and accurately. This capability is particularly valuable in the field of environmental monitoring, where AI algorithms can process satellite imagery, sensor data, and other sources of information to track changes in ecosystems, detect deforestation, and monitor air and water quality. By providing real-time insights into environmental conditions, AI enables policymakers, researchers, and conservationists to make informed decisions and take timely action to protect natural resources.

In agriculture, AI is revolutionizing the way we grow food by enabling precision farming techniques that maximize yields while minimizing environmental impact. Machine learning algorithms can analyze data from drones, sensors, and weather stations to optimize irrigation, fertilization, and pest control. This targeted approach reduces the use of water, chemicals, and energy, leading to more sustainable agricultural practices. Additionally, AI-powered predictive models can help farmers anticipate weather patterns and crop diseases, allowing them to adapt their strategies and improve resilience to climate change.

The energy sector is another area where AI is driving sustainability. By optimizing energy consumption and integrating renewable energy sources, AI can significantly reduce greenhouse gas emissions and promote cleaner energy systems. Smart grids, powered by AI, enable more efficient distribution of electricity by balancing supply and demand in real-time. This technology facilitates the integration of renewable energy sources, such as solar and wind, into the grid, reducing reliance on fossil fuels. Furthermore, AI can enhance energy efficiency in buildings by analyzing data from sensors and adjusting heating, cooling, and lighting systems to minimize energy use.

AI also plays a crucial role in waste management and recycling. Advanced algorithms can sort and categorize waste materials more accurately and efficiently than traditional methods, improving recycling rates and reducing landfill waste. AI-powered robots can identify and separate recyclable materials from mixed waste streams, ensuring that valuable resources are recovered and reused. By optimizing waste management processes, AI contributes to a circular economy, where resources are continuously reused and recycled, minimizing environmental impact.

Transportation is another sector where AI is making strides towards sustainability. Autonomous vehicles, powered by AI, have the potential to reduce traffic congestion, lower emissions, and improve fuel efficiency. By optimizing routes and driving patterns, AI can minimize energy consumption and reduce the carbon footprint of transportation systems. Additionally, AI can enhance public transportation systems by analyzing data on passenger demand and optimizing schedules and routes, making them more efficient and accessible.

In the realm of climate change mitigation, AI offers powerful tools for modeling and predicting the impacts of climate change. By analyzing historical climate data and simulating future scenarios, AI can help scientists and policymakers understand the potential consequences of different climate policies and interventions. This information is crucial for developing effective strategies to reduce greenhouse gas emissions and adapt to the changing climate. AI can also support carbon capture and storage technologies by optimizing processes and identifying the most effective methods for capturing and storing carbon dioxide.

AI's potential to drive sustainability extends beyond environmental applications. In the realm of social sustainability, AI can improve access to education, healthcare, and other essential services, particularly in underserved communities. By analyzing data on social and economic factors, AI can identify areas where resources are needed most and optimize the allocation of services. This targeted approach can help reduce inequalities and promote more equitable access to opportunities and resources.

Despite its potential, the integration of AI into sustainability efforts is not without challenges. Concerns about data privacy, algorithmic bias, and the environmental impact of AI technologies themselves must be addressed to ensure that AI contributes positively to sustainability goals. The energy consumption of AI systems, particularly those that rely on large-scale data processing and machine learning, is a significant consideration. Efforts to develop more energy-efficient AI technologies and promote the use of renewable energy in data centers are essential to mitigate these impacts.

Collaboration and interdisciplinary approaches are key to unlocking the full potential of AI for sustainability. By bringing together experts from diverse fields, including computer science, environmental science, engineering, and social sciences, we can develop innovative solutions that address complex sustainability challenges. Partnerships between governments, industries, academia, and civil society are crucial for fostering the development and deployment of AI technologies that promote sustainability.

Chapter 4

Sustainable Urban Development

Designing Eco-Friendly Cities

Urbanization has been a defining feature of modern society, with cities serving as hubs of economic activity, cultural exchange, and innovation. However, the rapid growth of urban areas has also led to significant environmental challenges, including pollution, resource depletion, and habitat destruction. Designing eco-friendly cities is an essential step towards creating sustainable urban environments that balance human needs with ecological preservation. By reimagining urban spaces, we can foster communities that thrive in harmony with nature.

One of the fundamental principles of eco-friendly city design is the integration of green spaces. Parks, gardens, and urban forests not only enhance the aesthetic appeal of cities but also provide critical ecosystem services. They improve air quality by absorbing pollutants and carbon dioxide, regulate temperatures through shading and evapotranspiration, and support biodiversity by providing habitats for various species. Green spaces also promote physical and mental well-being, offering residents opportunities for recreation, relaxation, and social interaction. To maximize these benefits, urban planners should prioritize the creation and maintenance of accessible green areas throughout the city.

Transportation is another key aspect of sustainable urban design. Traditional transportation systems, heavily reliant on fossil fuels, contribute significantly to air pollution and greenhouse gas emissions. To address this, eco-friendly cities must prioritize the development of efficient, low-emission transportation options. Public transit systems, such as buses, trams, and trains, should be expanded and optimized to provide reliable and convenient alternatives to private car use. Additionally, cities should invest in infrastructure that supports active transportation, such as cycling and walking. This includes the creation of dedicated bike lanes, pedestrian-friendly streets, and safe crossings. By reducing reliance on cars, cities can decrease emissions, alleviate traffic congestion, and improve public health.

The design of buildings and infrastructure plays a crucial role in the sustainability of urban environments. Eco-friendly cities should prioritize energy-efficient construction practices and materials that minimize environmental impact. This includes the use of renewable energy sources, such as solar panels and wind turbines, to power buildings and reduce dependence on fossil fuels. Incorporating passive design elements, such as natural lighting, ventilation, and insulation, can further enhance energy efficiency. Water conservation is another critical consideration, with strategies such as rainwater harvesting, greywater recycling, and low-flow fixtures helping to reduce water consumption.

Waste management is an essential component of sustainable city design. Eco-friendly cities should implement comprehensive waste reduction and recycling programs to minimize landfill use and promote a circular economy. This involves encouraging residents and businesses to reduce waste generation, separate recyclables, and compost organic materials. Advanced waste processing technologies, such as anaerobic digestion and waste-to-energy systems, can further enhance resource recovery and reduce environmental impact. Public education and awareness campaigns are vital to fostering a culture of sustainability and encouraging community participation in waste management efforts.

The concept of mixed-use development is central to the design of eco-friendly cities. By integrating residential, commercial, and recreational spaces within close proximity, cities can reduce the need for long commutes and promote walkability. Mixed-use neighborhoods encourage vibrant, diverse communities where residents can live, work, and play without relying on extensive transportation networks. This approach not only reduces emissions but also enhances quality of life by fostering social interaction and economic vitality.

Water management is a critical consideration in the design of sustainable urban environments. Cities must implement strategies to manage stormwater runoff, prevent flooding, and protect water quality. Green infrastructure, such as permeable pavements, green roofs, and rain gardens, can help absorb and filter rainwater, reducing the burden on drainage systems and minimizing pollution. Additionally, cities should prioritize the protection and restoration of natural water bodies, such as rivers, lakes, and wetlands, to maintain ecological balance and support biodiversity.

Community engagement is a vital aspect of designing eco-friendly cities. Residents should be actively involved in the planning and decision-making processes to ensure that urban development aligns with their needs and values. Public participation fosters a sense of ownership and responsibility, encouraging individuals to adopt sustainable practices and support environmental initiatives.

Cities can facilitate community involvement through workshops, forums, and collaborative projects that empower residents to contribute to the creation of sustainable urban spaces.

Technology and innovation play a significant role in the development of eco-friendly cities. Smart city technologies, such as sensors, data analytics, and the Internet of Things (IoT), can enhance urban efficiency and sustainability. These technologies enable real-time monitoring and management of resources, such as energy, water, and waste, allowing cities to optimize operations and reduce environmental impact. For example, smart grids can balance energy supply and demand, while intelligent transportation systems can improve traffic flow and reduce emissions. By leveraging technology, cities can create more responsive and adaptive urban environments.

The transition to eco-friendly cities requires collaboration and partnership among various stakeholders, including governments, businesses, and civil society. Policymakers must establish clear regulations and incentives to promote sustainable urban development, while businesses should adopt environmentally responsible practices and invest in green technologies. Civil society organizations can advocate for sustainability and engage communities in environmental initiatives. By working together, these stakeholders can drive the transformation of urban areas into sustainable, resilient, and livable spaces.

Green Architecture and Building Practices

The evolution of architecture has always been a reflection of society's values, technological advancements, and environmental awareness. In recent years, the growing concern for the planet's health has propelled green architecture and sustainable building practices to the forefront of the construction industry. This movement seeks to minimize the environmental impact of buildings while enhancing the well-being of their occupants. By embracing innovative design principles and materials, architects and builders can create structures that are not only aesthetically pleasing but also environmentally responsible.

One of the cornerstones of green architecture is energy efficiency. Buildings account for a significant portion of global energy consumption, making it imperative to reduce their energy footprint. This can be achieved through a combination of passive and active design strategies. Passive design focuses on optimizing the building's orientation, insulation, and natural ventilation to reduce the need for artificial heating and cooling. For instance, strategically placed windows can maximize natural light while minimizing heat gain, reducing the reliance on electric lighting and air conditioning. Active design, on the other hand, involves the integration of energy-efficient technologies such as solar panels, energy-efficient HVAC systems, and smart lighting controls. By harnessing renewable energy sources and optimizing energy use, buildings can significantly reduce their carbon emissions.

Water conservation is another critical aspect of sustainable building practices. With water scarcity becoming an increasingly pressing issue, architects must design buildings that use water efficiently. This can be achieved through the implementation of water-saving fixtures, such as low-flow toilets and faucets, as well as the use of greywater recycling systems. Rainwater harvesting is another effective strategy, allowing buildings to collect and store rainwater for non-potable uses such as irrigation and toilet flushing. By reducing water consumption and promoting the reuse of water, green buildings can contribute to the preservation of this vital resource.

The choice of materials plays a pivotal role in the sustainability of a building. Green architecture emphasizes the use of materials that are environmentally friendly, durable, and sourced responsibly. This includes materials with low embodied energy, such as recycled steel, reclaimed wood, and bamboo, which require less energy to produce and transport. Additionally, architects should prioritize materials that are non-toxic and promote healthy indoor air quality, such as low-VOC paints and finishes. The use of locally sourced materials can further reduce the environmental impact of construction by minimizing transportation emissions and supporting local economies.

The concept of biophilic design is gaining traction in the realm of green architecture. This approach seeks to connect building occupants with nature, enhancing their physical and mental well-being. Biophilic design elements can include the incorporation of natural materials, such as wood and stone, as well as the integration of indoor plants and green walls. Large windows and open spaces that provide views of the outdoors can also foster a sense of connection with nature. By creating environments that mimic natural settings, biophilic design can improve occupant satisfaction and productivity while reducing stress and promoting overall health.

Waste reduction is an essential component of sustainable building practices. The construction industry generates a significant amount of waste, much of which ends up in landfills. To address this issue, architects and builders should adopt strategies that minimize waste generation and promote recycling. This can include the use of modular construction techniques, which allow for precise material usage and reduce offcuts. Additionally, deconstruction, rather than demolition, can enable the salvage and reuse of building materials, reducing the need for new resources. By prioritizing waste reduction, the construction industry can contribute to a more sustainable future.

The integration of green roofs and living walls is an innovative approach to enhancing the sustainability of buildings. Green roofs, which are covered with vegetation, provide numerous environmental benefits, including improved insulation, stormwater management, and urban heat island mitigation. They also create habitats for wildlife and enhance the aesthetic appeal of urban areas. Living walls, or vertical gardens, can improve air quality, reduce noise pollution, and provide thermal insulation. These features not only contribute to the building's sustainability but also enhance the quality of life for its occupants and the surrounding community.

The role of technology in green architecture cannot be overstated. Advances in building technologies have enabled architects to design and construct more sustainable buildings than ever before. Building Information Modeling (BIM) allows for precise planning and simulation of a building's performance, enabling architects to optimize energy efficiency and resource use. Smart building systems, which use sensors and automation, can monitor and control energy consumption, lighting, and climate, ensuring optimal performance and comfort. By leveraging technology, architects can create buildings that are not only sustainable but also adaptable to changing environmental conditions.

Community engagement is a vital aspect of green architecture. Sustainable buildings should not only benefit their occupants but also contribute positively to the surrounding community. Architects should engage with local communities to understand their needs and values, ensuring that the building design aligns with the community's goals. This can include the incorporation of public spaces, such as parks and plazas, that promote social interaction and community cohesion. By fostering a sense of ownership and pride, green buildings can become integral parts of the community fabric.

The transition to green architecture and sustainable building practices requires a collaborative effort from architects, builders, policymakers, and the public. Governments can play a crucial role by establishing regulations and incentives that promote sustainable construction, such as green building certifications and tax credits. Builders and developers should prioritize sustainability in their projects, adopting best practices and innovative technologies.

The public can support green architecture by advocating for sustainable development and choosing environmentally responsible buildings.

Urban Transportation Solutions

Urban transportation has long been a cornerstone of city planning, shaping the way people move, interact, and live within metropolitan areas. As cities continue to grow, the demand for efficient, sustainable, and accessible transportation solutions becomes increasingly critical. The challenges of congestion, pollution, and limited space necessitate innovative approaches to urban mobility that prioritize the needs of both residents and the environment.

Public transportation systems are the backbone of urban mobility, offering a viable alternative to private car use. Buses, trains, and subways can move large numbers of people efficiently, reducing traffic congestion and lowering emissions. To enhance the effectiveness of public transit, cities must invest in expanding and modernizing their networks. This includes increasing the frequency and reliability of services, improving connectivity between different modes of transport, and ensuring accessibility for all users, including those with disabilities. By making public transportation more attractive and convenient, cities can encourage more residents to leave their cars at home.

The integration of technology into urban transportation systems has the potential to revolutionize the way people travel. Real-time data and mobile applications can provide users with up-to-date information on transit schedules, delays, and alternative routes, allowing them to make informed decisions about their journeys. Additionally, contactless payment systems and digital ticketing can streamline the boarding process, reducing wait times and enhancing the overall user experience. By leveraging technology, cities can create smarter, more responsive transportation networks that cater to the needs of their residents.

Cycling and walking are sustainable modes of transportation that offer numerous benefits, including improved public health, reduced emissions, and decreased traffic congestion. To promote these active forms of travel, cities must invest in the development of safe and accessible infrastructure. This includes the construction of dedicated bike lanes, pedestrian pathways, and secure bike parking facilities. Additionally, implementing traffic calming measures, such as reduced speed limits and pedestrian-friendly intersections, can enhance safety for cyclists and pedestrians. By creating an environment that supports active transportation, cities can encourage more residents to adopt these eco-friendly modes of travel.

Car-sharing and ride-sharing services have emerged as popular alternatives to private car ownership, offering flexible and cost-effective transportation options. These services can reduce the number of vehicles on the road, alleviating congestion and lowering emissions. To maximize the benefits of shared mobility, cities should establish policies and regulations that support the growth of these services while ensuring they complement existing public transportation networks. This may include designated pick-up and drop-off zones, incentives for electric vehicle use, and partnerships with transit agencies to provide first- and last-mile connectivity.

The rise of electric vehicles (EVs) presents an opportunity to reduce the environmental impact of urban transportation. EVs produce zero tailpipe emissions, contributing to improved air quality and reduced greenhouse gas emissions. To encourage the adoption of electric vehicles, cities must invest in the development of charging infrastructure, including public charging stations and incentives for home charging installations. Additionally, policies that promote the use of electric buses and taxis can further reduce the carbon footprint of urban transportation systems.

The concept of transit-oriented development (TOD) emphasizes the creation of compact, walkable communities centered around public transportation hubs. By concentrating residential, commercial, and recreational spaces near transit stations, TOD can reduce the need for car travel and promote sustainable urban growth. Cities can support transit-oriented development by implementing zoning policies that encourage higher-density development near transit corridors and providing incentives for mixed-use projects. By fostering vibrant, transit-friendly communities, cities can enhance the quality of life for their residents while reducing their reliance on private vehicles.

The implementation of congestion pricing is a strategy that can effectively manage traffic demand and reduce congestion in urban areas. By charging drivers a fee to enter high-traffic zones during peak hours, cities can encourage the use of alternative modes of transportation and reduce the number of vehicles on the road. The revenue generated from congestion pricing can be reinvested in public transportation infrastructure and services, further enhancing the overall transportation network. While congestion pricing may face initial resistance, its long-term benefits in terms of reduced traffic and improved air quality make it a valuable tool for urban transportation planning.

The development of autonomous vehicles holds the potential to transform urban transportation systems. Self-driving cars could improve traffic flow, reduce accidents, and provide mobility options for those unable to drive. However, the widespread adoption of autonomous vehicles also presents challenges, including potential increases in vehicle miles traveled and the need for updated infrastructure. Cities must proactively plan for the integration of autonomous vehicles, ensuring they complement existing transportation networks and contribute to sustainable urban mobility.

Community engagement is essential in the development of effective urban transportation solutions. By involving residents in the planning process, cities can better understand the needs and preferences of their communities, ensuring that transportation initiatives align with local priorities. Public consultations, workshops, and surveys can provide valuable insights and foster a sense of ownership and support for transportation projects. By prioritizing community input, cities can create transportation systems that are responsive to the needs of their residents and contribute to a more equitable and sustainable urban environment.

The future of urban transportation lies in the development of integrated, multimodal networks that prioritize sustainability, accessibility, and efficiency. By embracing a holistic approach to transportation planning, cities can create systems that meet the diverse needs of their residents while minimizing their environmental impact. This requires collaboration between government agencies, private sector partners, and the community to develop innovative solutions that address the challenges of urban mobility.

Community Engagement in Urban Planning

Community engagement in urban planning is a vital component of creating cities that reflect the needs and aspirations of their residents. As urban areas continue to expand and evolve, the importance of involving the community in the planning process cannot be overstated. By fostering a collaborative approach, cities can ensure that development projects are not only successful but also sustainable and inclusive.

The foundation of effective community engagement lies in open communication and transparency. Residents must be informed about upcoming projects, their potential impacts, and the opportunities available for participation. This can be achieved through various channels, such as public meetings, newsletters, social media, and dedicated websites. By providing clear and accessible information, cities can empower residents to become active participants in the planning process.

One of the most effective ways to engage the community is through public consultations. These events provide a platform for residents to voice their opinions, share their concerns, and contribute their ideas. Public consultations can take many forms, including town hall meetings, workshops, and focus groups. By creating a welcoming and inclusive environment, cities can encourage diverse participation and ensure that all voices are heard.

In addition to traditional consultation methods, cities can leverage technology to enhance community engagement. Online surveys, virtual town halls, and interactive mapping tools can reach a broader audience and provide valuable insights into community preferences and priorities. These digital platforms can also facilitate ongoing dialogue between residents and planners, fostering a sense of collaboration and shared responsibility.

To ensure meaningful community engagement, cities must prioritize inclusivity and equity. This means actively reaching out to underrepresented groups, such as low-income residents, minorities, and people with disabilities, and addressing any barriers to participation. Providing translation services, childcare, and accessible meeting locations can help ensure that all community members have the opportunity to contribute to the planning process.

The role of community organizations and local leaders is crucial in facilitating engagement. These groups often have deep connections within the community and can serve as valuable partners in the planning process. By collaborating with local organizations, cities can tap into existing networks and build trust with residents. This partnership can also help identify community champions who can advocate for the needs and priorities of their neighborhoods.

One of the key benefits of community engagement is the wealth of local knowledge and expertise that residents bring to the table. By tapping into this resource, cities can gain a deeper understanding of the unique challenges and opportunities within different neighborhoods. This local insight can inform more effective and context-sensitive planning decisions, ultimately leading to better outcomes for the community.

Community engagement also fosters a sense of ownership and pride among residents. When people feel that their voices have been heard and their contributions valued, they are more likely to support and invest in the success of a project. This sense of ownership can lead to increased civic participation and a stronger sense of community cohesion.

To maintain momentum and ensure ongoing engagement, cities must provide feedback to the community throughout the planning process. This includes sharing updates on project progress, explaining how community input has been incorporated, and addressing any concerns that may arise. By demonstrating a commitment to transparency and accountability, cities can build trust and strengthen relationships with their residents.

The challenges of community engagement should not be underestimated. Conflicting interests, limited resources, and time constraints can all pose obstacles to effective participation. However, by approaching these challenges with creativity and flexibility, cities can develop innovative solutions that meet the needs of their communities.

One approach to overcoming these challenges is to adopt a phased engagement strategy. This involves breaking the planning process into smaller, manageable stages, each with its own set of engagement activities. By focusing on specific issues or areas at each stage, cities can ensure that community input is targeted and relevant. This approach also allows for ongoing evaluation and adaptation, ensuring that engagement efforts remain effective and responsive to changing circumstances.

Another strategy is to create a community advisory board or task force. Composed of diverse representatives from the community, these groups can provide valuable input and guidance throughout the planning process. They can also serve as a bridge between residents and planners, facilitating communication and collaboration.

Ultimately, the success of community engagement in urban planning depends on a genuine commitment to collaboration and partnership. Cities must be willing to listen, learn, and adapt in response to community input. This requires a shift in mindset, from viewing residents as passive recipients of planning decisions to recognizing them as active partners in shaping the future of their cities.

By embracing community engagement as a core principle of urban planning, cities can create more resilient, inclusive, and vibrant communities. This collaborative approach not only leads to better planning outcomes but also strengthens the social fabric of the city, fostering a sense of belonging and shared purpose among residents.

Case Studies of Successful Sustainable Cities

The concept of sustainable cities has gained significant traction in recent years, as urban areas grapple with the challenges of rapid population growth, environmental degradation, and resource scarcity. By examining case studies of successful sustainable cities, we can glean valuable insights into the strategies and practices that have contributed to their achievements. These cities serve as beacons of innovation and resilience, demonstrating that sustainable urban development is not only possible but also beneficial for both people and the planet.

One of the most celebrated examples of a sustainable city is Copenhagen, Denmark. Renowned for its commitment to environmental stewardship, Copenhagen has set an ambitious goal to become carbon neutral by 2025. The city's success can be attributed to its comprehensive approach to sustainability, which encompasses transportation, energy, waste management, and urban design. A key component of Copenhagen's strategy is its emphasis on cycling as a primary mode of transportation. With an extensive network of bike lanes and infrastructure, the city has made cycling safe, convenient, and accessible for residents. This focus on sustainable mobility has not only reduced carbon emissions but also improved public health and quality of life.

In addition to its transportation initiatives, Copenhagen has invested heavily in renewable energy. The city is a pioneer in the use of wind power, with offshore wind farms supplying a significant portion of its electricity needs. Furthermore, Copenhagen has implemented district heating systems that utilize waste heat from power plants, reducing energy consumption and greenhouse gas emissions. These efforts have been complemented by stringent building codes and energy efficiency standards, ensuring that new developments adhere to the highest sustainability criteria.

Another exemplary sustainable city is Curitiba, Brazil. Often hailed as a model of urban planning, Curitiba has demonstrated that innovative solutions can transform a city into a more livable and environmentally friendly place. One of Curitiba's most notable achievements is its integrated public transportation system, which has been designed to prioritize efficiency and accessibility. The city's Bus Rapid Transit (BRT) system, introduced in the 1970s, has become a benchmark for cities worldwide. By providing fast, reliable, and affordable public transportation, Curitiba has successfully reduced traffic congestion and air pollution.

Curitiba's commitment to sustainability extends beyond transportation. The city has implemented a comprehensive waste management program that encourages recycling and waste reduction. Through initiatives such as the "Green Exchange" program, residents can exchange recyclable materials for fresh produce, promoting environmental awareness and social equity. Additionally, Curitiba has prioritized green spaces, with numerous parks and gardens integrated into the urban landscape. These green areas not only enhance the city's aesthetic appeal but also provide vital ecosystem services, such as air purification and flood mitigation.

Freiburg, Germany, is another city that has garnered international acclaim for its sustainability efforts. Known as the "Green City," Freiburg has embraced a holistic approach to urban development, focusing on renewable energy, sustainable transportation, and community engagement. The city's commitment to solar energy is particularly noteworthy, with numerous solar panels adorning rooftops and public buildings. Freiburg's dedication to renewable energy has been supported by strong policy frameworks and incentives, encouraging residents and businesses to invest in clean energy solutions.

Freiburg's transportation strategy emphasizes the importance of reducing car dependency. The city has developed an extensive network of cycling paths and pedestrian-friendly streets, making it easy for residents to choose sustainable modes of transportation. Public transportation is also a priority, with an efficient tram system that connects various parts of the city. By promoting sustainable mobility, Freiburg has successfully reduced its carbon footprint and improved air quality.

Community engagement is a cornerstone of Freiburg's sustainability efforts. The city actively involves residents in decision-making processes, fostering a sense of ownership and responsibility for the urban environment. This participatory approach has been instrumental in the success of various sustainability initiatives, as it ensures that projects align with the needs and values of the community.

Singapore is another city that has made significant strides in sustainability, despite its limited land area and high population density. The city-state has adopted a forward-thinking approach to urban planning, integrating green spaces and sustainable infrastructure into its development plans. Singapore's "Garden City" vision has resulted in a lush urban landscape, with parks, gardens, and green roofs scattered throughout the city. These green spaces not only enhance the city's aesthetic appeal but also provide essential ecosystem services, such as temperature regulation and biodiversity conservation.

Water management is a critical aspect of Singapore's sustainability strategy. The city has implemented an innovative water management system that includes rainwater harvesting, desalination, and water recycling. This comprehensive approach has enabled Singapore to achieve water self-sufficiency, reducing its reliance on external sources. Additionally, Singapore has invested in smart technology to optimize resource use and improve urban living conditions. The city's "Smart Nation" initiative leverages data and technology to enhance public services, transportation, and energy management.

Vancouver, Canada, is another city that has embraced sustainability as a guiding principle for urban development. The city's "Greenest City Action Plan" outlines ambitious targets for reducing carbon emissions, increasing green spaces, and promoting sustainable transportation. Vancouver's commitment to renewable energy is evident in its efforts to transition to a low-carbon economy. The city has invested in clean energy projects, such as wind, solar, and hydroelectric power, to reduce its reliance on fossil fuels.

Vancouver's transportation strategy prioritizes sustainable modes of travel, with an extensive network of cycling paths and public transit options. The city's focus on walkability and accessibility has made it easier for residents to choose environmentally friendly transportation options. Vancouver's dedication to sustainability extends to its building practices, with stringent green building standards in place to ensure that new developments are energy-efficient and environmentally responsible.